Jabriel & Jaiden Save New York City from Zombies

Jabriel & Jaiden
Save New York City from Zombies

Written by Jabriel Toscano
With Henry Charles
Illustrated by Eleonora Cali
Book cover/layout & logo designed by B. R. Graphix LLC

ISBN: 978-1-962325-90-5

THIS BOOK IS DEDICATED TO MY FAMILY

MY MOTHER, FATHER, BROTHER, AND TWO SISTERS.

Jabriel’s heart thumped loudly in his chest as he poured one of many combinations of mixtures into the beaker.

“How come the compound isn't responding?” exclaimed Jabriel.

The groaning from outside the bunker had now become deafening.

"They'll get inside, if we don't hurry. Higgs is now one of them, and we won't be able to get him back unless this works!"

Jabriel knew the process was sound, but he couldn't be sure if he had the correct mixtures with all the commotion.

"Ignore them right now. Help me check our work!"

"But the zomb—"

"This is our one and only chance! We're wasting time!"

The boys started going through the sequence, but Jaiden couldn't hear himself think because the bunker doors were rattling.

Thump. Thump. Thump.

"It's too late!" cried Jaiden. "We're screwed."

Urrrrhhhhh

Bang! Bang! Bang!

WHOMP. The monsters outside had finally gotten through the doors.

They had run out of time. But how?

Only two months ago, they couldn't wait for school to end. And now, a much more permanent end was racing toward them and the entire human race.

A few hours earlier:

On a warm June day, Uncle Higgs asked the brothers to stay after school.

"I can't believe that nut wants us to help him clean up his lab," whined 13-year-old Jaiden, as he wiped fresh beads of sweat from his forehead.

"Stop complaining," replied Jabriel. The 10-year-old had a soft spot for their eccentric uncle and hated when his older brother made fun of him.

"You're lucky you get to take chemistry with him. You get to see his genius in action with all those experiments. You should think twice before you call him crazy!"

Jaiden sighed and rolled his eyes. "Okay, whatever you say," he responded. "As long as he doesn't make me dust the funnels, he can tell us about whatever theory he wants."

Jabriel smiled. He loved his uncle's smartness and hoped to be just as smart as him someday. To the average person, the doomsday prepping would seem strange, but with everything that

Higgs knew, it only made sense that he would prepare for disasters too.

When the boys finally arrived at the lab at the back of the school, their uncle greeted them both with a bear hug.

"There are my favorite pupils!" beamed the scientist.

Higgs was dressed in tan pants with a red-and-white checkered dress shirt. He still had his white lab coat on, and a pair of dirty, old goggles sat atop his head. He stood there squeezing his two young nephews. Jabriel giggled, and Jaiden grunted. This made their uncle squeeze them more tightly and rock them back and forth before releasing them.

"Come on in!" exclaimed Higgs, motioning for them into the classroom.

However, something didn't feel right as the boys stepped inside.

"I'm confused," said Jaiden, with a furrowed brow. "I thought that we were here to clean up your lab. So, why is everything already cleaned up?"

Though Jabriel was simply happy to see his uncle, he, too, was beginning to wonder why they were there if they had nothing to tidy.

"Well," began Uncle Higgs, "You know that I've been prepping for a long time."

"Yes?" Said Jaiden, a bit worried about where the conversation was going.

"I believe that my preparations are about to be needed."

"What do you mean?" asked Jabriel, his eyes growing wide with curiosity.

"I've been keeping track of different stats from all over the world," continued the scientist, "And lately, I've noticed quite a few spikes."

"Spikes in what?" questioned Jaiden warily.

"There has been a large increase in liver consumption in a number of countries."

"Ew, gross!" squealed Jabriel.

"But why would that be alarming?" continued Jaiden.

Higgs took a deep breath, "it wouldn't be on its own," said the scientist, his face serious. "But there are other events that have occurred too."

"Like what?"

Higgs motioned for his nephews to sit in front of his desk while he brought up different files on his computer.

“Four weeks ago, a U.S. Navy ship was set to dock in the Dominican Republic but was denied entry at the port. The vessel had to reroute to Puerto Rico.”

“But why?” asked Jabriel, leaning forward in his seat.

“No explanation was ever reported in any newspaper, only that 20 crew members were transported to a local hospital with a serious illness.”

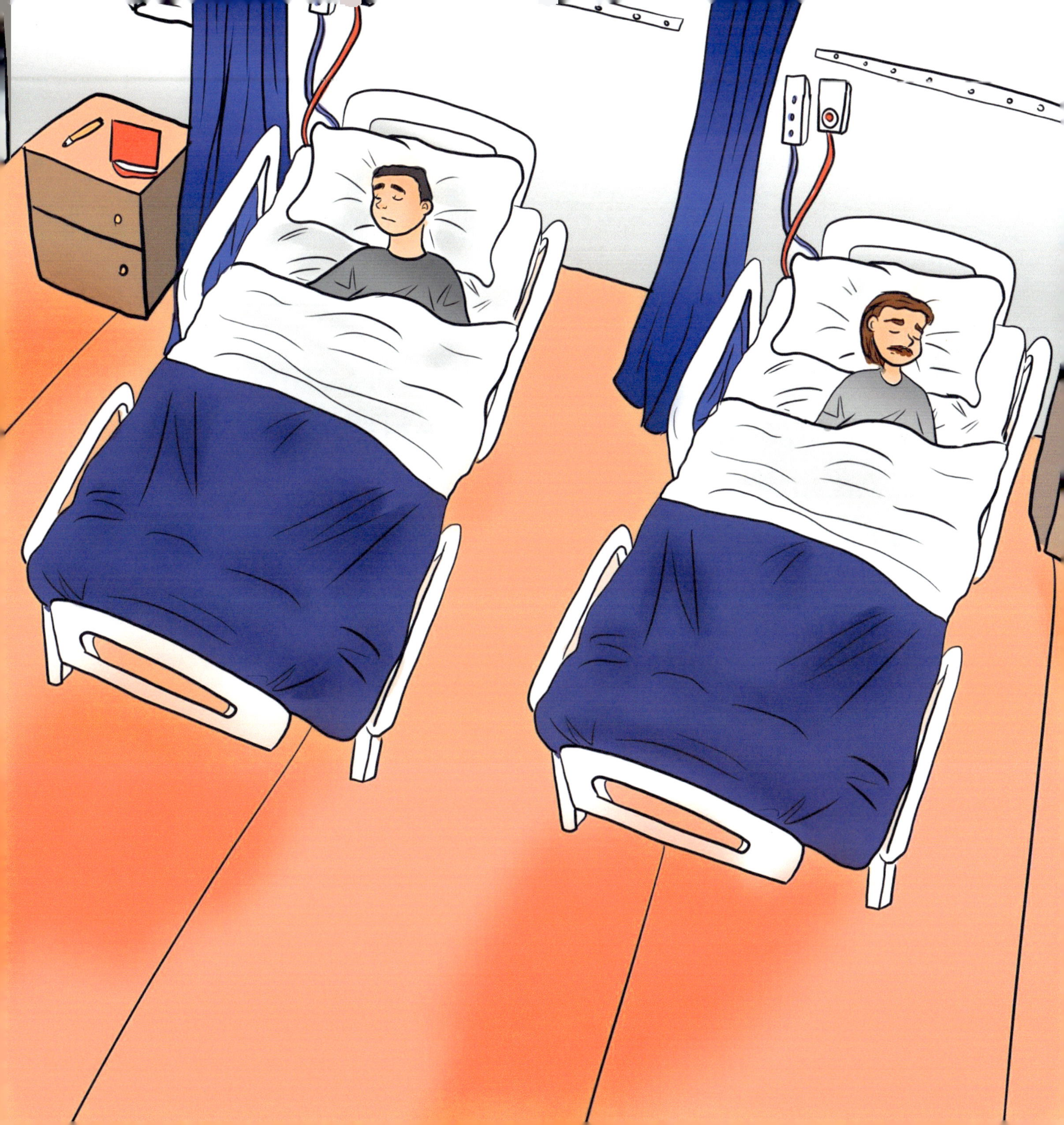

Jaiden was unimpressed: “So, some people were sick, so what’s the big deal?”

“An entire mall was also closed in Southeast Asia two weeks ago. Again, dozens of ambulances rushed people to a hospital, but no information about their condition appeared in the news.”

“Okay, so you are saying we’re dealing with a new kind of disease?” asked Jabriel, trying to impress his uncle.

"Exactly," said Higgs, straightening his lab coat. "However, I'm afraid it's unlike any other illness we've seen. Last week the police were called for a disturbance on the corner of East 75th Street and Park Ave...."

"That's nothing new!" squawked Jabriel. "Police are constantly dealing with crazy people."

"I agree," continued Uncle Higgs. "However, this woman bit two street vendors, and when the police arrived, she was doubled over and moaning something quite strange."

Jaiden was all ears, hunched forward in his seat. "What did she say?"

"The woman had her face close to her stomach and groaned like a wounded cat. The paramedics stated that she was calling for Brai—"

SCREECH. Someone had turned on the PA system.

"Attention, everyone, please," said a loud voice through the speakers. "Everyone, seek shelter immediately. I repeat, take shelter immediately."

The color drained from Higgs' face. Higgs grabbed a backpack, rummaged through it quickly, handed it to Jabriel and then fastened the straps on each of his shoulders. "We've got to get to Prospect Park."

"What? Why?" said both boys at the same time.

"It's zombies. We need to get to the bunker right away!"

The three ran through the grassy lawn that blanketed the park, approaching its western end. Sirens wailed as police cruisers and ambulances zipped along avenues to help those trapped by the undead.

"How could this have happened?" asked Jabriel, panting as he jogged beside his uncle.

"To put it simply," Higgs started, with heavy breaths, "a virus found a new home inside a human, and now anyone who gets the virus will crave—"

"Brains!" interjected Jaiden, his arms and feet swinging to keep up with his uncle's pace.

"Yes and no," replied Higgs. "At first, they only want liver, which is known to be nutritious. But eventually, the liver isn't enough, and they crave brains."

"Uncle Higgs?" said Jabriel, out of breath. "I need to stop, how far is it to the bunker?"

"It's just a little way," reassured Higgs.

"Okay, but we have to walk quickly then; the zombies are slow, but they are persistent. If they spot us, they will keep after us until they're restrained."

“I have to walk,” replied Jabriel. He wasn’t as fast as his brother, the star player on their school’s basketball team.

Despite the day's heat, the group continued quickly with determination across the park. The area would usually be bustling with families and friends dotting the grass. However, as the three trotted across the lawn toward Lookout Hill, there seemed to be no other souls beside them.

Or so it seemed.

“What’s that?” shrieked Jabriel, pointing to a grove of trees.

The group stopped dead in their tracks. A patch of vegetation rustled fifty yards away, and outstretched arms began to poke through, along with a series of strange moans.

"Run!" shouted Higgs.

The group took off.

"Head for those oak trees!" shouted Higgs, pointing to a grove across the field. As the group sprinted across, the moaning became louder.

"Oh no!" screamed Jaiden.

Behind the brush, ten yards from the bunker entrance, limbs emerged from the leaves.

"Come on! Come on!" Pleaded Jabriel as their uncle frantically entered the code for the bunker door.

"The door won't open fast enough!" Shouted the scientist, "You two slip in while the door is still opening, and I'll close it before the zombies can get inside."

"But Uncle Higgs! They'll catch you!" Protested Jabriel.

"Don't worry about me," reassured the scientist as he unfastened the backpack and handed it to his youngest nephew. "We need the formula. Now get going!"

"But-" Jabriel began.

"There are no buts! You'll find everything you need in my lab. Small zipper! Now go!"

With that, Higgs gently pushed each boy in through the narrow opening in the tree.

Urrrrhhhhh!

Jaiden turned around to look at his uncle one last time as Higgs pushed on his shoulders to help him through.With tears in his eyes, he saw green hands reach around his uncle as the scientist furiously entered the locking code.

"Be strong and work together, boys!" Said the uncle, struggling to hold the zombies back.

With a final press of a button, the door whirred closed. Jabriel locked eyes with his uncle in the last few rays of light. There was no fear in them. They were the last thing Jabriel saw as the hatch finally closed. He was now entirely in the dark.

"Come on!" a voice yelled, echoing in the distance. It was his brother. "We've got to get to the lab. Now!"

"But Uncle Higgs!" Said Jabriel, tears now streaming down his face.

"It's up to us now." Said the older brother. "If you want to save Higgs, start walking this way towards me."

"Okay," said the younger brother, his voice quivering.

And so, the boys moved slowly and carefully towards the lab. As they went, the muffled groans of the zombies grew quieter. But the thought of the creatures craving the boys’ flesh lingered in their minds.

The tunnel led about 30 yards before the boys entered the bunker.

Looking around, the two noticed all kinds of lab equipment surrounding them: beakers shaped like pears, test tubes coiled like snakes, vials with bubbling liquid, and other strange equipment surrounding the boys. However, they needed to find the recipe.

“How are we going to find the cure?” Asked Jabriel, his gaze fixed on one of the foaming beakers.

“We just need to start looking for something that resembles a recipe. You start at one end of the lab, and I’ll start from the other. Once you find a journal or something, let me know.”

The boys began their search. Jaiden bent down low and scoured the bottom shelves, but only bottles with different color chemical mixtures were there. Jabriel started rummaging through drawers, but despite all the tools they held, they did not find a single journal.

Jaiden began to feel discouraged.

"It's no use!" groaned the 13-year-old, "That crazy doomsday prepper has us on a wild goose chase."

"Stop talking about him like that!" Jabriel, standing tall to face his older brother, said, "He sacrificed himself for us! He wouldn't lie about this. The journal has to be somewhere!"

Jaiden looked at his younger brother. He stood there with clenched fists. The boy that had been crying for his uncle back in the tunnel was gone. In his place was one with a fierce determination.

Jaiden's tone softened. "You're right. I'm sorry. Let's stop and think for a minute," he said as he looked at his brother.

"We looked all over, and nothing," said Jaiden.

Urrhh!

Bang!

"Wait a minute. What did Higgs say to you?" said the elder sibling.

"I don't remember; he just handed me a backpack."

Urrrhhh

Bang! Bang!

The brothers are convinced. The zombies are right outside of the bunker.

"Zipper... Small zipper! Higgs said small zipper!"

The duo went for the backpack's small zipper compartment. Inside, a notepad with their favorite uncle's handwriting.

Higgs' Note:

"Hey boys, if you're reading this, that means we got separated, and you two are in the bunker. Based on the symptoms I have seen and read about, I have been working on a serum, and I may have a cure to reverse the effects of the zombies. Under my desk are two beakers, one blue and one red. You will need to mix them both in a large beaker on the Bunsen burner for a few minutes, then watch the mixture form a gas. The gas should reverse the effects. Higgs."

As Jaiden read the note from Higgs, Jabriel went in search of the red and blue mixtures. Moments later, Jabriel blurted, "I think I got it. There are so many beakers here with different colors. Many of them are blue and red." The youngest sibling carefully chose a blue and red bottled liquid, then carried the beakers over to the Bunsen burner.

"Hey, grab that empty beaker," his older brother directed. "Place it directly on top of the burner."

Jabriel followed his brother's instructions as Jaiden turned on the Bunsen burner.

"G'ahead," Jaiden said. His heart thumped loudly as his brother poured the last 50 grams of the concoction into the beaker.

"Why isn't the compound responding?" shouted Jabriel. The groaning from outside the bunker had now become deafening. "If we don't hurry, they'll get inside. Higgs is one of them by now, and we won't get him back unless this works!"

Jaiden knew that the process was sound, but he couldn't be sure he had chosen the correct mixtures with all the commotion.

"Ignore them right now. Help me check our work!"

"But the zomb—"

"This is our one and only chance! We're wasting time!"

The boys began to hear the bunker doors rattling.

Thump. Thump. Thump.

"It's too late!" cried Jabriel. "We're screwed."

Urrrrhhhhh

Bang! Bang! Bang!

WHOMP. The zombies outside have finally broken through the doors.

The brothers hid under a table.

"I guess this is it," said Jabriel.

"I'm sorry for what I said about Higgs," Jaiden admitted.

The zombies slowly wandered around the bunker.

Jaiden and Jabriel quietly crawled to a corner of the bunker. Suddenly, a thick white cloud began rapidly forming from the Bunsen burners beaker.

“Look, Jaiden, the mixture is working,” whispered his brother.

The entire bunker was engulfed in a thick white cloud, but the monsters kept getting closer and closer.

Urrrhhhhh!
Urrrrrrrrrrhhhhhhhhhh!

cough!
cough!

The cloud became so thick the boys couldn't see the ceiling. All they could see was the floor and the many feet of the monsters closing in on their position. Jaiden shielded his younger brother as the zombies came within inches of them.

Suddenly, a voice pierced from the thick cloud.

"Boys! I knew you two could do it!"

It was Higgs! The concoction worked. All the would-be monsters inside the bunker transformed back into their usual selves. Higgs kneeled and hugged his nephews.

"Looks like we're going to need more of this mixture," said Jaiden.

"Indeed, we do," said Higgs.

"Let's start now," Jaiden demanded.

"The ingredients are back in my apartment," said Higgs.

"Great. Let's go now!"
Jabriel said sternly.

The End?

From the Author

My name is Jabriel Toscano. I'm from Brooklyn, NY. I am 10 years old. I love to draw, write stories, play basketball and act. I love playing basketball with my dad and brother and would love to play professionally one day. My dad is a screenwriter and I've acted in two of his movies so far. This is my first story published, and I look forward to creating more. I hope you enjoy our story!

Watch my movie, streaming now on Tubi.

Follow me on Instagram **@brello1212**

Made in United States
Cleveland, OH
05 November 2026

47646339R00040